1. Greta Thunberg *No One Is Too Small to Make a Difference*
2. Naomi Klein *Hot Money*
3. Timothy Morton *All Art is Ecological*
4. George Monbiot *This Can't Be Happening*
5. Bill McKibben *An Idea Can Go Extinct*
6. Amitav Ghosh *Uncanny and Improbable Events*
7. Tim Flannery *A Warning from the Golden Toad*
8. Terry Tempest Williams *The Clan of One-Breasted Women*
9. Michael Pollan *Food Rules*
10. Robin Wall Kimmerer *The Democracy of Species*
11. Dai Qing *The Most Dammed Country in the World*
12. Wangari Maathai *The World We Once Lived In*
13. Jared Diamond *The Last Tree on Easter Island*
14. Wendell Berry *What I Stand for Is What I Stand On*
15. Edward O. Wilson *Every Species is a Masterpiece*
16. James Lovelock *We Belong to Gaia*
17. Masanobu Fukuoka *The Dragonfly Will Be the Messiah*
18. Arne Naess *There is No Point of No Return*
19. Rachel Carson *Man's War Against Nature*
20. Aldo Leopold *Think Like a Mountain*

This Can't Be Happening

GEORGE MONBIOT

PENGUIN BOOKS — GREEN IDEAS

PENGUIN BOOKS

UK | USA | Canada | Ireland | Australia
India | New Zealand | South Africa

Penguin Books is part of the Penguin Random House group of companies whose addresses can be found at global.penguinrandomhouse.com.

First published by the *Guardian* 2017, 2018, 2019
This selection published in Penguin Books 2021
001

'The Unseen World,' 'In Memoriam,' 'Deathly Silence,' 'Hopeless Realism, 'Intergenerational Theft, 'The Problem Is Capitalism' and 'Embarrassment of Riches' were first published in the *Guardian*. 'The New Political Story that Could Change Everything' was originally a talk delivered at TEDSummit 2019.

Set in 12.5/15pt Dante MT Std
Typeset by Jouve (UK), Milton Keynes
Printed and bound in Great Britain by Clays Ltd, Elcograf S.p.A.

A CIP catalogue record for this book is available from the British Library

The authorized representative in the EEA is Penguin Random House Ireland, Morrison Chambers, 32 Nassau Street, Dublin, D02 YH68

ISBN: 978-0-241-51463-4

www.greenpenguin.co.uk

Penguin Random House is committed to a sustainable future for our business, our readers and our planet. This book is made from Forest Stewardship Council® certified paper.

Contents

Contents

Foreword

For several months, I waged a lone campaign to persuade other parents waiting at the school gate to turn off their car engines. Wasn't it obvious that, by creating a toxic cloud around the school at pick-up time, they were harming their own children?

It wasn't obvious at all. My polite requests were met with polite bafflement. The most common response was 'why?' I found to my astonishment that most of those I spoke to seemed to be entirely unaware of air pollution. I don't think they were feigning ignorance. I believe they had never before been asked to make the connection between what came out of their exhausts and what goes into our lungs.

This might seem incredible, to those of us who have spent our lives engaged in such issues.

But it was a hard reminder that human survival is a niche interest.

The poisoning of the air and water, the collapse of wildlife and habitats, the heating of the atmosphere, the destruction of the living systems on which our lives depend might be the central, crucial facts of our age. But they have been deliberately pushed to the margins. They have been sidelined not passively, but actively and aggressively, by the billionaire press, the thinktanks that develop its talking points, the industries that finance the thinktanks and the governments that support the industries. All seek to distract and bamboozle, to deflect and delay, to impede our understanding of the greatest predicament humankind has ever faced.

Their task is easy. Engaging with the gathering collapse of our life support systems is mentally and morally challenging. It involves making connections that are seldom obvious. It means being prepared to hear bad news. Telling people what they want to hear – that their lives can carry on as before – is likely to win more friends than telling people that their apparently innocent activities are poisoning their children and destroying life on Earth.

How many of us really understand our situation? How many of us have grasped the full extent of what we face? Even those of us who engage with this crisis simultaneously know and shrink from knowing. Its speed and scale push the most fertile minds to their imaginative limits.

Previous generations have been forced to imagine existential crises: what if our crops fail? what if the plague arrives? what if raiders find our village? But, with the exception of those who had to consider the possibility of global nuclear war in the second half of the twentieth century, none have had to contemplate a realistic prospect of planetary collapse.

Arguably, nuclear war is easier to imagine than the predicament we now face. Were it to happen (and this threat, incidentally, has by no means passed), it would be a sudden and singular event. In one catastrophic spasm, the world would be transformed from home to wasteland. This annihilation, descending from the heavens, corresponds closely to religious visions of apocalypse. Horrible as it is, we are, in a way, pre-adapted to believe it.

But this – this is something different. We are asked to comprehend the gathering collapse of our habitable planet, caused not (or not only) by a cosmic misjudgement on the part of a few men with godlike powers, but by the tiny, incremental things we do. Believing it requires, above all, a leap of moral imagination.

The moral compass needed to guide us through this crisis has had to be constructed almost from scratch. We need to understand that we can inflict great harm on others, without harbouring any such intent. We need to connect our humdrum activities on one side of the world (generally in the rich nations) to catastrophic effects on the lives of people on the other side (generally in the poorer nations). We need to see that the way we live can destroy the life chances of those who are not yet born.

Many of our core beliefs drive us in the wrong direction. Concepts such as 'progress', 'wealth creation' and 'self-improvement' come to mean, in the context of environmental collapse, the opposite of what they meant before. Culturally and morally, we are pre-adapted not to accept what environmental science is telling us.

Eventually, I gave up trying to persuade the other parents. I could see them thinking, though they were too polite to say it, 'Hold off! Unhand me, grey-beard loon!' It had no lasting, discernible impact. It was yet another reminder of the futility of acting alone.

We act effectively through movements, coming together to demand systemic change. These movements are now gathering, almost as rapidly as the crises we contest.

Clearly, we need to demand practical changes: new technologies, new economies, new ways of living. But we also, and perhaps primarily, need to do something deeper: to overcome the ignorance the billionaire press has manufactured, to wake our friends from the stupor of consumption, to break through the barrier of disbelief and provoke a new moral imagination. I hope this collection might, in a small way, help.

August 2020

The Unseen World

What you see is not what others see. We inhabit parallel worlds of perception, bounded by our interests and experience. What is obvious to some is invisible to others. I might find myself standing, transfixed, by the roadside, watching a sparrowhawk hunting among the bushes, astonished that other people could ignore it. But they might just as well be wondering how I could have failed to notice the new V6 Pentastar Sahara that just drove past.

As the psychologist Richard Wiseman points out, 'At any one moment, your eyes and brain only have the processing power to look at a very small part of your surroundings . . . your brain quickly identifies what it considers to be the most significant aspects of your surroundings, and focuses almost all of its attention on these elements.' Everything else remains unseen.

Our selective blindness is lethal to the living world. Joni Mitchell's claim that 'you don't know what you've got till it's gone' is, sadly, untrue: our collective memory is wiped clean by ecological loss. One of the most important concepts defining our relationship to the living world is Shifting Baseline Syndrome, coined by the fisheries biologist Daniel Pauly. The people of every generation perceive the state of the ecosystems they encountered in their childhood as normal and natural. When wildlife is depleted, we might notice the loss, but we are unaware that the baseline by which we judge the decline is in fact a state of extreme depletion.

So we forget that the default state of almost all ecosystems – on land and at sea – is domination by a mega fauna. We are unaware that there is something deeply weird about British waters: namely that they are not thronged with great whales, vast shoals of bluefin tuna, two-metre cod and halibut the size of doors, as they were until a few centuries ago. We are unaware that the absence of elephants, rhinos, lions, scimitar cats, hyenas and hippos, which lived in this country during the last interglacial period (when the climate was almost identical to today's) is also an artefact of human activity.

And the erosion continues. Few people younger than me know that it was once normal to see fields white with mushrooms, or rivers black with eels at the autumn equinox. I can picture a moment when the birds stop singing, and people wake up and make breakfast and go to work without noticing that anything has changed.

Conversely, the darkness in which we live ensures that we don't know what we have, even while it exists. The BBC's *Blue Planet II* revealed the complex social lives and remarkable intelligences of species we treat as nothing but seafood (a point it failed to drive home, in its profoundly disappointing final episode). If we were aware of the destruction we commission with our routine purchases of fish, would we not radically change our buying habits? But the infrastructure of marketing and media helps us not to see, not to think, not to connect our spots of perception to create a moral worldview upon which we can act.

Most people subconsciously collaborate in this evasion. It protects them from either grief or cognitive dissonance. To be aware of the wonder and enchantment of the world, its astonishing creatures and complex interactions, and to be aware simultaneously of the remarkably rapid destruction of almost every living system, is to take on a burden of grief that is almost unbearable. This is what the great conservationist Aldo Leopold meant when he wrote that 'One of the penalties of an ecological education is that one lives alone in a world of wounds.'

In June this year, a powerful light – 125 watts, to be precise – was shone into a corner of my own darkness. Two naturalists from Flanders, Bart van Camp and Rollin Verlinde, asked if they could come to our tiny urban garden and set up a light trap. The results were a revelation.

I had come to see the garden – despite our best efforts – as almost dead: butterflies and beetles are rare sights here. But when Bart and Rollin showed us the moths they had caught, I realized that what we see does not equate to what there is. There are fifty-nine species of butterfly in the UK, but 2,500 species of moth, and our failure to apprehend the ecology of darkness limits our understanding of the living world.

When they opened the trap, I was astonished by the range and beauty of their catch. There were pink and olive elephant hawkmoths; a pine hawkmoth, feathered and ashy; a buff arches, patterned and gilded like the back of a barn owl; flame moths in polished brass; the yellow kites of swallowtail moths; common emeralds the colour of a northern sea, with streaks of foam; grey daggers; a pebble prominent; heart and darts; coronets; riband waves; willow beauties; an elder pearl; small magpie; double-striped pug; rosy tabby: the names testify to a rich relationship between these creatures and those who love them.

Altogether, there were 217 moths of fifty species. This, they told me, was roughly what they had expected to find. Twenty-five years ago, there would have been far more. A food web is collapsing, probably through a combination of pesticides, habitat destruction and light pollution, and we are scarcely aware of its existence.

Moths evolved around 190 million years ago. By comparison, butterflies are a recent development, diverging 140 million years later. Most explanations for this split focus on the spread of flowering plants. But might it have more to do with the fact that bats developed echolocation at roughly that time? Could the diurnal butterfly have been a response to a deadly adaptation by the nocturnal moth's main predator?

Every summer night, an unseen drama unfolds over our gardens, as moths, whose ears are tuned to the echo-locating sounds bats make, drop like stones out of the sky to avoid predation. Some tiger moths have evolved to jam bat sonar, by producing ultra-sonic clicks of their own. We destroy the wonders of the unseen world before we appreciate them.

That morning I became a better naturalist,

and a better conservationist. I began to look more closely, to seek the unseen, to consider what lies beneath. And to realize just how much there is to lose.

December 2017

In Memoriam

It felt as disorientating as forgetting my pin number. I stared at the caterpillar, unable to attach a name to it. I don't think my mental powers are fading: I still possess an eerie capacity to recall facts and figures and memorize long screeds of text. This is a specific loss. As a child and young adult, I delighted in being able to identify almost any wild plant or animal. And now this ability has gone. It has shrivelled from disuse: I can no longer identify them because I can no longer find them.

Perhaps this forgetfulness is protective. I have been averting my eyes. Because I cannot bear to see what we have done to nature, I no longer see nature itself. Otherwise, the speed of loss would be unendurable. The collapse can be witnessed from one year to the next. The swift decline of the swift (down 25 per cent in five years) is marked by the loss of the wild screams that, until very recently, filled the skies above my house. My ambition to see the seabird colonies of the Shetlands and St Kilda has been replaced by the intention never to visit those islands during the breeding season: I could not bear to see the empty cliffs, whose populations have crashed by some 90 per cent this century.

I have lived long enough to witness the vanishing of wild mammals, butterflies, mayflies, songbirds and fish that I once feared my grandchildren would experience: it has all happened faster than even the pessimists predicted. Walking in the countryside or snorkelling in the sea is now as painful to me as an art lover would find her visits to a gallery, if on every occasion another Old Master had been cut from its frame.

The cause of this acceleration is no mystery. The United Nations reports that our use of natural resources has tripled in forty years. The great expansion of mining, logging, meat production and industrial fishing is cleansing the planet of its wild places and natural wonders.

What economists proclaim as progress, ecologists recognize as ruin.

This is what has driven the quadrupling of oceanic dead zones since 1950; the 'biological annihilation' represented by the astonishing collapse of vertebrate populations; the rush to carve up the last intact forests; the vanishing of coral reefs, glaciers and sea ice; the shrinkage of lakes, the drainage of wetlands. The living world is dying of consumption.

We have a fatal weakness: a failure to perceive incremental change. As natural systems shift from one state to another, we almost immediately forget what we have lost. I have to make a determined effort to remember what I saw in my youth. Could it really be true that every patch of nettles, at this time of year, was reamed with caterpillar holes? That flycatchers were so common I scarcely gave them a second glance? That moths of all shapes and sizes crowded the windows on summer nights?

Others seem oblivious. When I have criticized current practice, farmers have sent me images of verdant monocultures of perennial rye grass, with the message: 'Look at this and try telling me we don't look after nature.' It's green, but it's about as ecologically rich as an airport runway. One of my readers, Michael Groves, records the shift he has seen in the field beside his house, where the grass, that used to be cut for hay, is now cut for silage. Watching the cutters being driven at great speed across the field, he realized that any remaining wildlife would be shredded. Soon afterwards, he saw a roe deer standing in the mown grass. She stayed there throughout the day and the following night. When he went to investigate, he found her fawn, its legs amputated. 'I felt sickened, angry and powerless . . . How long had it taken to die?' That 'grass-fed meat' the magazines and restaurants fetishize? This is the reality.

When our memories are wiped as clean as the land, we fail to demand its restoration. Our forgetting is a gift to industrial lobby groups and the governments that serve them. Over the past few months, I have been told repeatedly that the environment secretary, Michael Gove, gets it. I have said so myself: he genuinely seems to understand what the problems are and what needs to be done. Unfortunately, he doesn't do it.

He cannot be blamed for all of the fiascos to which he has put his name. The 25-year plan for nature was, it seems, gutted by the Prime Minister's Office. The environmental watchdog he proposed was defanged by the Treasury (it has subsequently been lent some dentures by Parliament). Other failures are his own work. In response to lobbying from sheep farmers, he has allowed ravens, a highly intelligent and long-lived species just beginning to recover from centuries of persecution, to be killed once more. There are 24 million sheep in this country and 7,400 pairs of ravens. Why must all other species give way to the white plague?

Responding to complaints that most of our national parks are wildlife deserts, Gove set up a commission to review them. But governments choose their conclusions in advance, through the appointments they make. A more dismal, backward-looking and uninspiring panel would be hard to find: not one of its members, as far as

I can tell, has expressed a desire for significant change in our national parks, and most of them, if their past statements are anything to go by, are determined to keep them in their sheep-wrecked and grouse-trashed state.

Now the lobbyists demand a New Zealand settlement for farming after Brexit: deregulated, upscaled, hostile to both wildlife and the human eye. If they get their way, no landscape, however treasured, will be safe from broiler sheds and mega-dairy units, no river protected from run-off and pollution, no songbird saved from local extinction. The merger between Bayer and Monsanto brings together the manufacturer of the world's most lethal pesticides with the manufacturer of the world's most lethal herbicides. Already the concentrated power of these behemoths is a hazard to democracy; together they threaten both political and ecological disaster. Labour's environment team have scarcely a word to say about any of it. Similarly, the big conservation groups, as usual, have gone missing in inaction.

We forget even our own histories. We fail to

recall, for example, that the Dower Report, published in 1945, envisaged wilder national parks than we now possess, and that the conservation white paper the government issued in 1947 called for the kind of large-scale protection that is considered edgy and innovative today. Remembering is a radical act.

That caterpillar, by the way, was a six-spot burnet: the larva of a stunning iridescent black and pink moth that once populated my neighbourhood and my mind. I will not allow myself to forget again: I will work to recover the knowledge I have lost. For I now see that without the power of memory, we cannot hope to defend the world we love.

June 2018

Deathly Silence

We're getting there, aren't we? We're making the transition towards an all-electric future. We can now leave fossil fuels in the ground and thwart climate breakdown. Or so you might imagine if you follow the technology news.

So how come oil production, for the first time in history, is about to hit 100 million barrels per day? How come the oil industry expects demand to climb until the 2030s? How is it that in Germany, whose energy transition (Energiewende) was supposed to be a model for the world, protesters are being beaten up by police as they try to defend the 12,000-year-old Hambacher Forest from an opencast mine extracting lignite: the dirtiest form of coal? Why have investments in Canadian tar sands – the dirtiest source of oil – doubled in the past year?

The answer is growth. There might be more electric vehicles on the world's roads, but there are also more internal combustion engines. There might be more bicycles, but there are also more planes. It doesn't matter how many good things we do; preventing climate breakdown means ceasing to do bad things. Given that economic growth, in nations that are already rich enough to meet the needs of all, requires the growth of pointless consumption, it is hard to see how it can ever be decoupled from the assault on the living planet.

When a low-carbon industry expands within a growing economy, the money it generates stimulates high-carbon industry. Anyone who works in this field knows environmental entrepreneurs, eco-consultants and green business managers who use their earnings to pay for holidays in distant parts of the world, and the flights required to get there. Electric vehicles have driven a new resource rush, particularly for lithium, that is already polluting rivers and trashing precious wild places. Clean growth is as much of an oxymoron as clean coal. But making this obvious statement in public life is treated as political suicide.

The Labour Party's new environment policy, published this week, rightly argues that 'our current economic model is threatening the foundations on which human wellbeing depends.' It recognizes that ecological collapse cannot be prevented through consumer choice or corporate social responsibility. The response to our greatest predicament must be determined by scientific research, and planned, coordinated and led by government. It pledges 'to meet the Paris Agreement goal of limiting global temperature rises to no more than 1.5°C'.

But, as almost everyone does, it ignores the fundamental problem. Beyond a certain point, economic growth, the force that lifted people out of poverty, cured deprivation, squalor and disease, tips us back into those conditions. To judge by the devastation climate breakdown is wreaking, we appear already to have reached this point.

The contradiction is most obvious when the policy document discusses airports (an issue on which the party is divided). Labour guarantees that any airport expansion 'adheres to our tests' on climate change. But airport expansion is incompatible with its climate commitments. Even if aircraft emissions are capped at 2005 levels, by 2050 they will account for half the nation's carbon budget, if the UK is not to contribute to more than 1.5°C of global warming. If airports grow, they will swallow even more of the budget.

Airport expansion is highly regressive, offending the principles of justice and equity that Labour exists to uphold. Regardless of the availability and cost of flights, they are used disproportionately by the rich, as these are the people with the business meetings in New York, the second homes in Tuscany and the money to pay for winter holidays in the sun. Yet the impacts – noise, pollution and climate breakdown – are visited disproportionately on the poor.

I recognize that challenging our least contested ideologies – growth and consumerism – is a tough call. But in New Zealand, it is beginning to happen. Jacinda Ardern, the Labour

Prime Minister, says, 'It will no longer be good enough to say a policy is successful because it increases GDP if . . . it also degrades the physical environment.' How this translates into policy, and whether her party will resolve its own contradictions, remains to be determined.

No politician can act without support. If we want political parties to address these issues, we too must start addressing them. We cannot rely on the media to do it for us. A report by the research group Media Matters found that total coverage of climate change across five US news networks (ABC, CBS, NBC, Fox and PBS) amounted to 260 minutes in 2017 – a little over four hours. Almost all of it was a facet of the Trump psychodrama – Will he pull out of the Paris Accord? What's he gone and done this time? – rather than the treatment of climate chaos in its own right. There was scarcely a mention of the link between climate breakdown and the multiple unnatural disasters the US suffered that year, of new findings in climate science or the impacts of new pipelines or coal mines. I cannot find a comparable recent study in the UK. I suspect the situation here is a little better, but not a lot.

The worst denial is not the claim that this existential crisis isn't happening. It is the failure to talk about it at all. Not talking about our greatest predicament, even as it starts to bite, requires a constant and determined effort. Taken as a whole (of course there are exceptions), the media is a threat to humanity. It claims to speak on our behalf. But it either speaks against us or does not speak at all.

So what do we do? We talk. As the climate writer Joe Romm argued on the US website ThinkProgress earlier this year, a crucial factor in the remarkable shift in attitudes towards LGBT people was the determination of activists to break the silence. They overcame social embarrassment to broach issues that other people found uncomfortable. We need, Romm said, to do the same for climate breakdown.

A recent survey suggests that 65 per cent of Americans rarely or never discuss it with friends or family, while only one in five hear people they know mention the subject at least once a month. Like the media, we subconsciously invest great psychological effort into not discussing an issue that threatens almost every aspect of our lives.

Let's be embarrassing. Let's break the silence, however uncomfortable it makes us and others feel. Let's talk about the great unmentionables: not just climate breakdown, but also growth and consumerism. Let's create the political space in which well-intentioned parties can act. Let us talk a better world into being.

September 2018

A Good Start

The launch of Extinction Rebellion, Parliament Square, London

Why are we here?

We are here because for decades scarcely anyone else has been doing what we are now standing up (or sitting down), to do.

I've been in this business for thirty-three years altogether, at the BBC, then as a freelancer, at the *Guardian*, writing about, campaigning against, the destruction of our living planet.

And every single year, I have heard people say, 'Well, the government hasn't done enough, but it's a good start.'

Every year there's been a good start.

Every year, there's been a beginning. But never a middle or an end to this story.

Why not?

Why, when so many people are so desperate to see action?

It is because, though we claim to live in a democracy, in many respects our system of government resembles a plutocracy.

Where the voice of the people should be heard, the money of the City and the fossil fuel industry, and the farming lobby, and the fishing industry, and the auto manufacturers, and the airlines lobby counts instead.

We are not heard because they are heard.

We are excluded because they are included.

The will of the people is not done.

Parliament will not do this for us. For thirty-three years I have waited, and for thirty-three years it has not happened.

Corporations will not do this for us.

And, I'm sorry to say, the big NGOs will not do this for us either. Where we need mass action, where we need radicalism, instead we've seen half measures.

Well, this, my friends, is where the full measures begin.

We are here to defend the living world that others have not defended.

We are here to support the life support systems that defend us.

We are here to keep fossil fuels in the ground.

To get the mass livestock industry off the land.

To stop the ripping apart of the oceans by super-trawlers and other forms of destruction pushed by profits.

To shut down airport space.

To get a sensible transport policy that puts people and planet first, and fossil fuels last.

Above all, to dethrone economic growth as the objective of government.

To replace it with the only thing that counts, which is the wellbeing of all the species on Earth. The wellbeing of humanity and of the other creatures with which we share this wonderful planet. The only planet known to support intelligent life – though how 'intelligent' it is remains to be demonstrated.

We want not just to protect what we already have, but to restore what we have lost.

To bring the trees back to the barren hills.

To let the rivers run freely once more.

To allow the seas to surge with whales and dolphins again.

To see the great profusion, the wonders of life that our ancestors knew and our great-grandchildren could know again.

We are here, my friends, for the once and future planet. No one else will deliver it for us. No one is left but us.

Where does the EXTINCTION REBELLION begin? Here.

With whom does the EXTINCTION REBELLION begin? With us.

We have waited long enough. We are waiting no longer.

The only time people know it's serious is when people are prepared to sacrifice their liberty in defence of their beliefs.

We are those people.

We put our hands up in defence of Mother Earth.

We put our hands up in defence of humanity.

We put our hands up against Extinction, and for life.

October 2018

Hopeless Realism

It was a moment of the kind that changes lives. At a press conference held by Extinction Rebellion last week, two of us journalists pressed the activists on whether their aims were realistic. They have called, for example, for carbon emissions in the UK to be reduced to net zero by 2025. Wouldn't it be better, we asked, to pursue some intermediate aims?

A young woman called Lizia Woolf stepped forward. She hadn't spoken before, and I hadn't really noticed her, but the passion, grief and fury of her response were utterly compelling. 'What is it that you are asking me as a twenty-year-old to face and to accept about my future and my life? . . . This is an emergency – we are facing extinction. When you ask questions like that, what is it you want me to feel?' We had no answer.

Softer aims might be politically realistic, but they are physically unrealistic. Only shifts commensurate with the scale of our existential crises have any prospect of averting them. Hopeless realism, tinkering at the edges of the problem, got us into this mess. It will not get us out.

Public figures talk and act as if environmental change will be linear and gradual. But the Earth's systems are highly complex, and complex systems do not respond to pressure in linear ways. When these systems interact (because the world's atmosphere, oceans, land surface and lifeforms do not sit placidly within the boxes that make study more convenient), their reactions to change become highly unpredictable. Small perturbations can ramify wildly. Tipping points are likely to remain invisible until we have passed them. We could see changes of state so abrupt and profound that no continuity can be safely assumed.

Only one of the many life support systems on which we depend – soils, aquifers, rainfall, ice, the pattern of winds and currents, pollinators, biological abundance and diversity – need fail for everything to slide. For example, when Arctic sea ice melts beyond a certain point, the positive feedbacks this triggers (such as darker water absorbing more heat, melting permafrost releasing methane, shifts in the polar vortex) could render runaway climate breakdown unstoppable. When the Younger Dryas period ended 11,600 years ago, Greenland ice cores reveal local temperatures rising 10°C within a decade.

I don't believe that such a collapse is yet inevitable, or that a commensurate response is either technically or economically impossible. When the US joined the Second World War in 1941, it replaced a civilian economy with a military economy within months. As Jack Doyle records in his book *Taken for a Ride*, 'In one year, General Motors developed, tooled, and completely built from scratch 1,000 Avenger and 1,000 Wildcat aircraft . . . Barely a year after Pontiac received a Navy contract to build antishipping missiles, the company began delivering the completed product to carrier squadrons around the world.' And this was before advanced information technology made everything faster.

The problem is political. A fascinating analysis by the social science professor Kevin MacKay contends that oligarchy has been a more fundamental cause of the collapse of civilizations than social complexity or energy demand. Oligarchic control, he argues, thwarts rational decision-making, because the short-term interests of the elite are radically different to the long-term interests of society. This explains why past civilizations have collapsed 'despite possessing the cultural and technological know-how needed to resolve their crises'. Economic elites, which benefit from social dysfunction, block the necessary solutions.

The oligarchic control of wealth, politics, media and public discourse explains the comprehensive institutional failure now pushing us towards disaster. Think of Trump and his cabinet of multi-millionaires, the influence of the Koch brothers, the Murdoch empire and its massive contribution to climate science denial, the oil and motor companies whose lobbying prevents a faster shift to new technologies.

It is not just governments that have failed to respond, though they have failed spectacularly. Public sector broadcasters have deliberately and systematically shut down environmental coverage, while allowing the opaquely funded lobbyists that masquerade as think-tanks to shape public discourse and deny what we face. Academics, afraid to upset their funders and colleagues, have bitten their lips. Even the bodies that claim to be addressing our predicament remain locked within destructive frameworks.

For example, last Wednesday I attended a meeting about environmental breakdown at the Institute for Public Policy Research. Many of the people in the room seemed to understand that continued economic growth is incompatible with sustaining the Earth's systems. As the author Jason Hickel points out, a decoupling of rising GDP from global resource use has not happened and will not happen. While 50 billion tonnes of resources used per year is roughly the limit the Earth's systems can tolerate, the world is already consuming 70 billion tonnes. Business as usual, at current rates of economic growth, will ensure that this rises to 180 billion tonnes by 2050. Maximum resource efficiency, coupled with massive carbon taxes and some pretty optimistic assumptions, would reduce this to 95 billion tonnes: still way beyond environmental limits. A study taking account of the rebound effect (efficiency leads to further resource use) raises the estimate to 132 billion tonnes. Green growth, as members of the Institute appear to accept, is physically impossible.

On the same day, the same Institute announced a major new economics prize for 'ambitious proposals to achieve a step-change improvement in the growth rate'. It wants ideas that will enable economic growth rates in the UK at least to double. The announcement was accompanied by the usual blah about sustainability, but none of the judges of the prize has a discernible record of environmental interest.

Those to whom we look for solutions trundle on as if nothing has changed. They continue to behave as if the accumulating evidence has no purchase on their minds. Decades of institutional failure ensure that only 'unrealistic' proposals – the repurposing of economic life, with immediate effect – now have a realistic chance of stopping the planetary death spiral. And only those who stand outside the failed institutions can lead this effort.

Two tasks need to be performed simultaneously: throwing ourselves at the possibility of averting collapse, as Extinction Rebellion is doing, slight though this possibility may appear. And preparing ourselves for the likely failure of these efforts, terrifying as this prospect is. Both tasks require a complete revision of our relationship with the living planet. Because we cannot save ourselves without contesting oligarchic control, the fight for democracy and justice and the fight against environmental breakdown are one and the same. Do not allow those who have caused this crisis to define the limits of political action. Do not allow those whose magical thinking got us into this mess to tell us what can and cannot be done.

November 2018

Intergenerational Theft

The young people taking to the streets are right: their future is being stolen. The economy is an environmental pyramid scheme, dumping its liabilities on the young and the unborn. Its current growth depends on intergenerational theft.

At the heart of capitalism is a vast and scarcely examined assumption: you are entitled to as great a share of the world's resources as your money can buy. You can purchase as much land, as much atmospheric space, as many minerals, as much meat and fish as you can afford, regardless of who might be deprived. If you can pay for them, you can own entire mountain ranges and fertile plains. You can burn as much fuel as you like. Every pound or dollar secures a certain right over the world's natural wealth. But why? What just principle equates the numbers in your bank account with a right to own the fabric of the Earth? Most people I ask are completely stumped by this question.

The standard justification goes back to John Locke's *Second Treatise of Government*, published in 1689. Locke claimed that you acquire a right to own natural wealth by mixing your labour with it: the fruit you pick, the minerals you dig and the land you till become your exclusive property, because you put the work in.

This argument was developed in the 18th century by the jurist William Blackstone, whose books were immensely influential in England, America and elsewhere. He contended that a man's right to 'sole and despotic dominion' over land was established by the person who first occupied it, to produce food. This right could then be exchanged for money. This is the underlying rationale for the great pyramid scheme. And it makes no sense.

For a start, it assumes a Year Zero. At this arbitrary point, a person could step onto a piece of land, mix their labour with it, and claim it as theirs. Locke used America as an example of the blank slate on which people could establish their rights. But the land (as Blackstone admitted) became a blank slate only through the extermination of those who lived there.

Not only could the colonist erase all prior rights, he could also erase all future rights. By mixing your labour with the land once, you and your descendants acquire the right to it in perpetuity, until you decide to sell it. You thereby prevent all future claimants from gaining natural wealth by the same means.

Worse still, according to Locke, 'your' labour includes the labour of those who work for you. But why should the people who do the work not be the ones who acquire the rights? It's comprehensible only when you realize that by 'man', Locke means not all humankind, but European men of property. Those who worked for them had no such rights. What this meant, in the late 17th century, was that large-scale land rights could be justified, under his system, only by the ownership of slaves. Inadvertently perhaps, Locke produced a charter for the human rights of slave holders.

Even if these objections could somehow be dismissed, what is it about labour that magically turns anything it touches into private property? Why not establish your right to natural wealth by peeing on it? The arguments defending our economic system are flimsy and preposterous. Peel them away, and you see that the whole structure is founded on looting: looting from other people, looting from other nations, looting from other species and looting from the future.

Yet, on the grounds of these absurdities, the rich arrogate to themselves the right to buy the natural wealth on which others depend. Locke cautioned that his justification works only if 'there is enough, and as good, left in common for others'. Today, whether you are talking about land, the atmosphere, living systems, rich mineral lodes or most other forms of natural wealth, it is clear that there is not 'enough, and as good' left in common.

Everything we take for ourselves we take from someone else.

You can tweak this system. You can seek to modify it. But you cannot make it just.

So what should take its place? It seems to me that the founding principle of any just system is that those who are not yet alive will, when they are born, have the same rights as those who are alive today. At first sight, this doesn't seem to change anything: the first article of the Universal Declaration states that 'All human beings are born free and equal in dignity and rights.' But this statement is almost meaningless, because there is nothing in the declaration insisting that one generation cannot steal from the next. The missing article might look like this: 'Every generation shall have an equal right to the enjoyment of natural wealth.'

This principle is hard to dispute, but it seems to change everything. Immediately, it tells us that no renewable resource should be used beyond its rate of replenishment. No non-renewable resource should be used that cannot be fully recycled and reused. This leads inexorably towards two major shifts: a circular economy from which materials are never lost, and the end of fossil fuel combustion.

But what of the Earth itself? In this densely populated world, all land ownership necessarily precludes ownership by others. Article 17 of the Universal Declaration is self-contradictory. It says, 'Everyone has the right to own property.' But because it places no limit on the amount one person can possess, it ensures that everyone does not have this right. I would change it to this: 'Everyone has the right to use property without infringing the rights of others to use property.' The implication is that everyone born today would acquire an equal right of use, or would need to be compensated for their exclusion. One way of implementing this is through major land taxes, paid into a sovereign wealth fund. It would alter and restrict the concept of ownership, and ensure that economies tended towards distribution, rather than concentration.

These simple suggestions raise a thousand

questions. I don't have all the answers. But such issues should be the subject of lively conversations everywhere. Preventing environmental breakdown and systemic collapse means challenging our deepest and least-examined beliefs.

March 2019

The Problem Is Capitalism

For most of my adult life, I've railed against 'corporate capitalism', 'consumer capitalism' and 'crony capitalism'. It took me a long time to see that the problem is not the adjective, but the noun.

While some people have rejected capitalism gladly and swiftly, I've done so slowly and reluctantly. Part of the reason was that I could see no clear alternative: unlike some anti-capitalists, I have never been an enthusiast for state communism. I was also inhibited by its religious status. To say 'capitalism is failing' in the 21st century is like saying 'God is dead' in the 19th. It is secular blasphemy. It requires a degree of self-confidence I did not possess.

But as I've grown older, I've come to recognize two things. First, that it is the system, rather than any variant of the system, which drives us inexorably towards disaster. Second, that you do not have to produce a definitive alternative to say that capitalism is failing. The statement stands in its own right. But it also demands another, and different, effort to develop a new system.

Capitalism's failures arise from two of its defining elements. The first is perpetual growth. Economic growth is the aggregate effect of the quest to accumulate capital and extract profit. Capitalism collapses without growth, yet perpetual growth on a finite planet leads inexorably to environmental calamity.

Those who defend capitalism argue that, as consumption switches from goods to services, economic growth can be decoupled from the use of material resources. Last week, a paper in the journal *New Political Economy* by Jason Hickel and Giorgos Kallis examined this premise. They found that while some relative decoupling took place in the 20th century (material resource consumption grew, but not as quickly as economic growth), in the 21st there has been a re-coupling: rising resource consumption has so far matched or exceeded the rate of economic growth. The absolute decoupling needed to avert environmental catastrophe (a reduction in material resource use) has never been achieved, and appears impossible while economic growth continues. Green growth is an illusion.

A system based on perpetual growth cannot function without peripheries and externalities. There must always be an extraction zone, from which materials are taken without full payment, and a disposal zone, where costs are dumped in the form of waste and pollution. As the scale of economic activity increases, until capitalism affects everything from the atmosphere to the deep ocean floor, the entire planet becomes a sacrifice zone: we all inhabit the periphery of the profit-making machine.

This drives us towards cataclysm on such a scale that most people have no means of imagining it. The threatened collapse of our life support systems is bigger by far than war, famine, pestilence or economic crisis, though it is likely to incorporate all four. Societies can recover from these apocalyptic events, but not from the loss of soil, an abundant biosphere and a habitable climate.

The second defining element is the bizarre assumption that a person is entitled to as great a share of the world's natural wealth as their money can buy. This seizure of common goods causes three further dislocations. First, the scramble for exclusive control of non-reproducible assets, which implies either violence or legislative truncations of other people's rights. Second, the immiseration of other people by an economy based on looting across both space and time. Third, the translation of economic power into political power, as control over essential resources leads to control over the social relations that surround them.

In *The New York Times* on Sunday, the Nobel economist Joseph Stiglitz sought to distinguish between good capitalism, that he called 'wealth creation', and bad capitalism, that he called 'wealth grabbing' (extracting rent). I understand his distinction, but from the environmental point of view, wealth creation is wealth grabbing. Economic growth, intrinsically linked to the increasing use of material resources, means seizing natural wealth from both living systems and future generations.

To point to such problems is to invite a barrage of accusations, many of which are based on this premise: capitalism has rescued hundreds of millions of people from poverty – now you want to impoverish them again. It is true that capitalism, and the economic growth it drives, has radically improved the prosperity of vast numbers of people, while simultaneously destroying the prosperity of many others: those whose land, labour and resources were seized to fuel growth elsewhere. Much of the wealth of the rich nations was – and is – built on slavery and colonial expropriation.

Like coal, capitalism has brought many benefits. But, like coal, it now causes more harm than good. Just as we have found means of generating useful energy that are better and less damaging than coal, so we need to find means of generating human wellbeing that are better and less damaging than capitalism.

There is no going back: the alternative to capitalism is neither feudalism nor state communism. Soviet communism had more in common with capitalism than the advocates of either system would care to admit. Both systems are (or were) obsessed with generating economic growth. Both are willing to inflict astonishing levels of harm in pursuit of this and other ends. Both promised a future in which we would need to work for only a few hours a week, but instead demand endless, brutal labour. Both are dehumanizing. Both are absolutist, insisting that theirs and theirs alone is the one true God.

So what does a better system look like? I don't have a complete answer, and I don't believe any one person does. But I think I see a rough framework emerging. Part of it is provided by the ecological civilization proposed by Jeremy Lent, one of the greatest thinkers of our age. Other elements come from Kate Raworth's doughnut economics and the environmental thinking of Naomi Klein, Amitav Ghosh, Angaangaq

Angakkorsuaq, Raj Patel and Bill McKibben. Others still are derived from the notion of 'private sufficiency, public luxury'.

I believe our task is to identify the best proposals from many different thinkers and shape them into a coherent alternative. Because no economic system is only an economic system, but intrudes into every aspect of our lives, we need many minds from various disciplines – economic, environmental, political, cultural, social and logistical – working collaboratively to create a better way of organizing ourselves that meets our needs without destroying our home.

Our choice comes down to this. Do we stop life to allow capitalism to continue, or stop capitalism to allow life to continue?

March 2019

The New Political Story that Could Change Everything

TED Summit 2019
EICC, Edinburgh

Do you feel trapped in a broken economic model? A model that's trashing the living world and threatens the lives of our descendants? A model that excludes billions of people while making a handful unimaginably rich? That sorts us into winners and losers, and then blames the losers for their misfortune? Welcome to neoliberalism, the zombie doctrine that never seems to die, however comprehensively it is discredited.

You might have imagined that the financial crisis of 2008 would have led to the collapse of neoliberalism. After all, it exposed its central features, which were deregulating business and finance, tearing down public protections, throwing us into extreme competition with each other, as being, well, just a little bit flawed. And intellectually, it did collapse. But still, it dominates our lives.

Why? Well, I believe the answer is that we have not yet produced a new story with which to replace it.

Stories are the means by which we navigate the world. They allow us to interpret its complex and contradictory signals. When we want to make sense of something, the sense we seek is not scientific sense but narrative fidelity. Does what we are hearing reflect the way that we expect humans and the world to behave? Does it hang together? Does it progress as a story should progress?

We are creatures of narrative, and a string of facts and figures, however important facts and figures are, has no power to displace a persuasive story. The only thing that can replace a story is a story. You cannot take away someone's story without giving them a new one.

It's not just stories in general that we are attuned to, but particular narrative structures.

There are a number of basic plots that we use again and again, and in politics there is one basic plot which turns out to be tremendously powerful, and I call this 'the restoration story'. It goes as follows:

Disorder afflicts the land, caused by powerful and nefarious forces working against the interests of humanity. But the hero will revolt against this disorder, fight those powerful forces, against the odds overthrow them and restore harmony to the land.

You've heard this story before. It's the Bible story. It's the Harry Potter story. It's the Lord of the Rings story. It's the Narnia story. But it's also the story that has accompanied almost every political and religious transformation going back millennia. In fact, we could go as far as to say that without a powerful new restoration story, a political and religious transformation might not be able to happen. It's that important.

After laissez-faire economics triggered the Great Depression, John Maynard Keynes sat down to write a new economics, and what he did was to tell a restoration story, which went something like this:

Disorder afflicts the land! Caused by the powerful and nefarious forces of the economic elite, which have captured the world's wealth. But the hero of the story, the enabling state, supported by working class and middle-class people, will contest that disorder, will fight those powerful forces by redistributing wealth, and through spending public money on public goods will generate income and jobs, restoring harmony to the land.

Now, like all good restoration stories, this one resonated across the political spectrum. Democrats and Republicans, Labour and Conservatives, left and right all became, broadly, Keynesian.

Then, when Keynesianism ran into trouble in the 1970s, the neoliberals, people like Friedrich Hayek and Milton Friedman, came forward with their new restoration story, and it went something like this:

Disorder afflicts the land! Caused by the powerful and nefarious forces of the overmighty state, whose collectivizing tendencies crush freedom and individualism and opportunity.

But the hero of the story, the entrepreneur, will fight those powerful forces, roll back the state, and, through creating wealth and opportunity, restore harmony to the land.

And that story also resonated across the political spectrum. Republicans and Democrats, Conservatives and Labour, they all became, broadly, neoliberal. Opposite stories with an identical narrative structure.

Then, in 2008, the neoliberal story fell apart, and its opponents came forward with . . . nothing. No new restoration story! The best they had to offer was a watered-down neoliberalism or a microwaved Keynesianism.

And that is why we're stuck. Without that new story, we are stuck with the old failed story that keeps on failing.

Despair is the state we fall into when our imagination fails. When we have no story that explains the present and describes the future, hope evaporates.

Political failure is at heart a failure of imagination. Without a restoration story that can tell us where we need to go, nothing is going to change. But with such a restoration story, almost everything can change. The story we need to tell is a story which will appeal to as wide a range of people as possible, crossing political fault lines. It should resonate with deep needs and desires. It should be simple and intelligible, and it should be grounded in reality.

I admit that all of this sounds like a tall order. But I believe that in Western nations, there is a story like this waiting to be told.

Over the past few years, there's been a fascinating convergence of findings in several different sciences, in psychology and anthropology and neuroscience and evolutionary biology, and they all tell us something pretty amazing: that human beings have a massive capacity for altruism. Sure, we all have a bit of selfishness and greed inside us, but in most people, those are not our dominant values. And we also turn out to be the supreme cooperators. We survived the African savannas, despite being weaker and slower than our predators and most of our prey, by an amazing ability to engage in mutual aid, and that urge to cooperate has been hardwired into our minds through natural selection. These are the central, crucial facts about humankind: our amazing altruism and cooperation.

But something has gone horribly wrong. Disorder afflicts the land.

Our good nature has been thwarted by several forces, but I think the most powerful of them is the dominant political narrative of our times, which tells us that we should live in extreme individualism and competition with each other. It pushes us to fight each other, to fear and mistrust each other. It atomizes society. It weakens the social bonds that make our lives worth living. And into that vacuum grow these violent, intolerant forces. We are a society of altruists, but we are governed by psychopaths.

But it doesn't have to be like this, because we have an incredible capacity for togetherness and belonging. By invoking that capacity, we can recover those amazing components of our humanity: our altruism and cooperation. Where there is atomization, we can build a thriving civic life with a rich participatory culture. Where we find ourselves crushed between market and state, we can build an economics that respects both people and planet. And we can create this economics around that great neglected sphere, the commons.

The commons is neither market nor state, capitalism nor communism, but it consists of three main elements: a particular resource; a particular community that manages that resource; and the rules and negotiations the community develops to manage it. Think of community broadband or community energy cooperatives or the shared land for growing fruit and vegetables that in Britain we call allotments. A commons can't be sold, it can't be given away, and its benefits are shared equally among the members of the community.

Where we have been ignored and exploited, we can revive our politics. We can recover democracy from the people who have captured it. We can use new rules and methods of elections to ensure that financial power never trumps democratic power again.

Representative democracy should be tempered by participatory democracy, so that we can refine our political choices, and this choice should be exercised as much as possible at the local level. If something can be decided locally, it shouldn't be determined nationally.

I call all this the politics of belonging.

I think it has the potential to appeal across quite a wide range of people, and the reason for this is that among the very few values that both left and right share are belonging and community. We might mean slightly different things by them, but at least we start with some language in common. In fact, you can see a lot of politics as being a search for belonging. Even fascists seek community, albeit a frighteningly homogenous community where everyone looks the same and wears the same uniform and chants the same slogans.

What we need to create is a community based on bridging networks, not bonding networks. A bonding network brings together people from a homogenous group, whereas a bridging network brings together people from different groups. And my belief is that if we create sufficiently rich and vibrant bridging communities, we can thwart the urge for people to burrow into the security of a homogenous bonding community, defending themselves against the OTHER.

So in summary, our new story could go something like this:

Disorder afflicts the land!

Caused by the powerful and nefarious forces of people who say there's no such thing as society, who tell us that our highest purpose in life is to fight like stray dogs over a dustbin. But the heroes of the story – us – will revolt against this disorder. We will fight those nefarious forces by building rich, engaging, inclusive and generous communities, and, in doing so, we will restore harmony to the land.

Now, whether or not you feel this is the right story, I hope you'll agree that we need one. We need a new restoration story, which tells us why we're in this mess and tells us how to get out of that mess. And this story, if we tell it right, will infect the minds of people across the political spectrum.

Our task is to tell the story that lights the path to a better world.

July 2019

Embarrassment of Riches

It is not quite true that behind every great fortune lies a great crime. Musicians and novelists, for example, can become extremely rich by giving other people pleasure. But it does appear to be universally true that in front of every great fortune lies a great crime. Immense wealth translates automatically into immense environmental impacts, regardless of the intentions of those who possess it. The very wealthy, almost as a matter of definition, are committing ecocide.

A few weeks ago, I received a letter from a worker at a British private airport. 'I see things that really shouldn't be happening in 2019,' he wrote. Every day he sees Global 7000 jets, Gulfstream 650s and even Boeing 737s take off from the airport carrying a single passenger, mostly flying to Russia and the US. The private Boeing

737s, built to take 174 seats, are filled at the airport with around 32,000 litres of fuel. That's as much fossil energy as a small African town might use in a year.

Where are these single passengers going? Perhaps to visit one of their super-homes, constructed and run at vast environmental cost, or to take a trip on their super-yacht, which might burn 500 litres of diesel per hour just ticking over and is built and furnished with rare materials, extracted at the expense of stunning places.

Perhaps we shouldn't be surprised to learn that when Google convened a meeting of the rich and famous at the Verdura resort in Sicily this July to discuss climate breakdown, its delegates arrived in 114 private jets and a fleet of mega-yachts, and drove around the island in super-cars. Even when they mean well, the ultra-rich cannot help trashing the living world.

A series of research papers shows that income is by far the most important determinant of environmental impact. It doesn't matter how green you think you are. If you have surplus money, you spend it. The only form of consumption that's clearly and positively correlated with good environmental intentions is diet: people who see themselves as green tend to eat less meat and more organic vegetables. But attitudes have little bearing on the amount of transport fuel, home energy and other materials you consume. Money conquers all.

The disastrous effects of spending power are compounded by the psychological impacts of being wealthy. Plenty of studies show that the richer you are, the less you are able to connect with other people. Wealth suppresses empathy. One paper reveals that drivers in expensive cars are less likely to stop for people using pedestrian crossings than drivers in cheap cars. Another shows that rich people are less able than poorer people to feel compassion towards children with cancer.

Though they are disproportionately responsible for our environmental crises, the rich will be hurt least and last by planetary disaster, while the poor are hurt first and worst. The richer people are, the research suggests, the less such knowledge is likely to trouble them.

Another issue is that wealth limits the perspectives of even the best-intentioned people. This week Bill Gates argued in an interview with the *Financial Times* that divesting (ditching stocks) from fossil fuels is a waste of time. It would be better, he claimed, to pour money into disruptive new technologies with lower emissions. Of course we need new technologies. But he has missed the crucial point: in seeking to prevent climate breakdown, what counts is not what you do but what you stop doing. It doesn't matter how many solar panels you install if you don't simultaneously shut down coal and gas burners. Unless existing fossil fuel plants are retired before the end of their lives, and all exploration and development of new fossil fuels reserves is cancelled, there is little chance of preventing more than 1.5°C of global heating.

But this requires structural change, which involves political intervention as well as technological innovation: anathema to Silicon Valley billionaires. It demands an acknowledgement that money is not a magic wand that makes all the bad stuff go away.

On Friday, I'll be joining the global climate strike, in which adults will stand with the young people whose call to action has resonated around the world. As a freelancer, I've been wondering who I'm striking against. Myself? Yes: one aspect of myself, at least. Perhaps the most radical thing we can now do is to limit our material aspirations. The assumption on which governments and economists operate is that everyone strives to maximize their wealth. If we succeed in this task, we inevitably demolish our life support systems. Were the poor to live like the rich, and the rich to live like the oligarchs, we would destroy everything. The continued pursuit of wealth, in a world that has enough already (albeit very poorly distributed), is a formula for mass destitution.

A meaningful strike in defence of the living world is, in part, a strike against the desire to raise our incomes and accumulate wealth: a desire shaped, more than we are probably aware, by dominant social and economic narratives. I see myself as striking in support of a radical and disturbing concept: Enough. Individually and collectively, it is time to decide what enough looks like, and how to know when we've achieved it.

There's a name for this approach, coined by the Belgian philosopher Ingrid Robeyns: limitarianism. Robeyns argues that there should be an upper limit to the amount of income and wealth a person can amass. Just as we recognize a poverty line, below which no one should fall, we should recognize a riches line, above which no one should rise. This call for a levelling down is perhaps the most blasphemous idea in contemporary discourse.

But her arguments are sound. Surplus money allows some people to exercise inordinate power over others, in the workplace, in politics, and above all in the capture, use and destruction of natural wealth. If everyone is to flourish, we cannot afford the rich. Nor can we afford our own aspirations, which the culture of wealth maximization encourages.

The grim truth is that the rich are able to live as they do only because others are poor: there is neither the physical nor ecological space for everyone to pursue private luxury. Instead we should strive for private sufficiency, public luxury. Life on Earth depends on moderation.

September 2019

The Weird Are Destined to Change the World

Extinction Rebellion,
Lambeth Bridge, London

For decades they called people like us mad. And when we were few in number, that hurt.

But now that there are millions of us, it doesn't hurt any more.

For decades they called people like us weird. And when we were few in number, that hurt.

But now that there are millions of us, it doesn't hurt any more.

We own this label, and we wear it with pride.

Do you know the etymology of the word 'weird'? It comes from the old English 'wyrd'. Do you know what wyrd meant? It meant destiny.

The weird are destined to change the world.

It's no coincidence that the person who has transformed the entire global conversation about climate breakdown has Asperger's syndrome. Throughout human evolution our survival has depended on many different ways of looking at the world. A group of hominids who all saw things the same way would have been less likely to survive than a group who saw things in different ways.

Not only has Greta Thunberg transformed the global conversation about climate breakdown. She has also transformed the global conversation about neurodiversity. And we honour her for both these blessings.

But we are induced by schooling and politics to see the world from one perspective only. It recognizes only one kind of intelligence: linear, analytical, focused on work and jobs and capital. It fails to recognize many others: spatial, navigational, creative, caring, social, holistic.

It is this narrowness, this restrictive definition of sanity, that drives us towards true madness. The madness that is destroying our life support systems. Our schooling fails us individually, and fails us collectively.

One of the extraordinary aspects of this

movement is the way we have schooled ourselves. We have seen for ourselves what is important, and learnt that it is not self-enrichment. It is not the accumulation of stuff. It is not the pursuit of status.

We have come to see that what is salient in the media and the commercial world is not important, and what is important is not salient. We're here to make the important salient. To turn what the science is telling us into the central fact of public life.

We are not here to negotiate. There is nothing to negotiate about. You cannot negotiate with physics. You cannot decide to suspend the laws of thermodynamics. There is no halfway point at which politicians can meet us. Because either they respond appropriately to what the science is saying, or they do not.

We are constantly told that we have to be politically realistic. That our demands are not politically realistic and can therefore be ignored.

As we have seen over the past three years in Britain, political realism is a somewhat flexible concept. What is deemed impossible one month suddenly becomes possible, even inevitable, the next. But there is no such flexibility with scientific realism.

Do you know the old Situationist slogan, from fifty years ago? 'Be realistic, demand the impossible.' It is time to reclaim and repurpose it:

'Be scientifically realistic. Demand the politically impossible.'

And until they give us what they deem to be politically impossible, we will keep demanding. We will return again and again, in all our glorious madness and weirdness, to demand that those who see themselves as normal and sane recognize that true sanity, in the face of climate breakdown and ecological breakdown, demands a completely new system.

A system that functions within planetary limits. A system that sits within Earth's systems, rather than bursting through them. A system on a human scale and a natural scale that respects both humanity and the rest of the living world.

And for this we rise in rebellion. We rise in a

rebellion built on the most embarrassing word in the English language: love. We rebel in love for each other, in love for the living world, in love for the better people we can become.

We rebel against extinction and we rebel for life.

October 2019

We Can Stop the First Great Extermination

The Launch of Animal Rebellion, Smithfield Market, London

I have spent my life learning and unlearning. We cannot navigate the world in which we live without unlearning some of its deepest myths.

One of the most enduring stories is the pastoral myth: the idea that the city is corrupt and evil, while the shepherd and his flock are innocent and pure. This story is thousands of years old. It recurs in so many cultures that I wonder if it originated among the nomads of central Asia, who occupied both India and south-eastern Europe. It was told by the ancient Greeks and by the scribes who wrote the Old Testament. By the Roman poets and Elizabethan writers – by Spenser, Marlowe and Shakespeare. It was revived by the Romantics, and lives on to this day in hundreds of television programmes. If the BBC were any keener on sheep, it would be illegal.

A version of the pastoral myth appears in some of the first books that any child encounters: the farmyard tales that occupy our moral imagination from our earliest glimmerings of consciousness. The story they tell is remarkably consistent: a farm with one cow, one pig, one horse, one sheep, one chicken and one rosy-cheeked farmer, living together in bucolic harmony. There is, of course, no hint of why animals might be kept on a farm, or what fate awaits them. These farms bear no relation to any farm I have ever seen, yet somehow the old, old story still governs our perception of where our food comes from. It can take a lifetime to unlearn. I suspect many people never do unlearn it.

It took me long enough, even though I worked on livestock farms when I was a teenager. I didn't challenge the pastoral myth until I moved to Mid Wales. I lived in the Dyfi Valley, between Snowdonia and the Cambrian mountains.

At first, I was excited. I could walk all day in any direction, and scarcely cross a road or see a house. But soon, my excitement gave way to puzzlement. My puzzlement gave way to disappointment. My disappointment gave way to despair. The land was not just empty of people, it was empty of wildlife.

Above around 200 metres, there were no trees. There were no birds. I could walk all day and see two crows and a pipit if I was lucky. I could get down on my hands and knees in the middle of summer and find no insects in the sward. It was a dead land.

Only one kind of animal was in abundance: sheep.

Slowly, it began to dawn on me that the two things might be connected. I started to understand that the landscape was empty because of the sheep. I discovered that sheep selectively browse out tree seedlings: when the trees that covered these mountains died, there were no young ones to replace them. The entire ecosystem, across hundreds of years, had been wiped out by livestock.

Then I realized that what I was seeing in Mid Wales was almost universal: across Britain, sheep have reduced once thriving uplands to bare monocultures. Obvious as it now was, I had failed to see it because I had not yet unlearnt what I was taught as a child.

Farmers insist that they are feeding the nation. But is this really true? I discovered, through my own research, as there are no official figures, that sheep farming in the uplands occupies roughly four million hectares in the UK. That's more land than we use for growing grain. It's twenty-four times as much land as we use for growing fruit and vegetables. Yet sheep in this country, upland and lowland, provide just 1 per cent of our food.

This, in other words, is a classic example of agricultural sprawl: huge areas of land used to yield a tiny amount of food. All over the world, livestock grazing, while producing just a small proportion of our food, wipes out wildlife and habitats on an enormous scale.

Well-meaning TV chefs and food writers tell us we should switch from intensively reared meat to pasture-fed meat. But all this does is to swap a system that causes astonishing cruelty for one that causes astonishing destruction. If everyone did as they suggest, we would quickly run out of planet. There simply isn't enough land.

Free-range chicken farming can also be extremely damaging to the living world. Take a look at what is happening to the beautiful River Wye and its tributaries on the Welsh border. They are being turned, at staggering speed, into open sewers. In high summer they stink – literally. The primary reason is the proliferation of free-range chicken farms in the area. Chickens kept in large numbers outdoors lay down a scorching carpet of reactive phosphate. Rain washes it into the rivers, causing algal blooms and wiping out much of their wildlife.

We often hear about 'wildlife–human conflicts'. But almost all of them, in reality, are conflicts between wildlife and livestock. Why are badgers being persecuted in this country? Why are wild boar being slaughtered across Europe? Why are wolves, coyotes and pumas massacred in North America? Because of their impact on livestock.

Some livestock farmers say that, through 'regenerative farming', they are mimicking nature. It is true that, in some respects, their practices are better than conventional livestock farming, though their most arresting claims, especially concerning the amount of carbon they store, are grossly exaggerated. But they don't mimic nature: they reduce it to a parody of its former abundance and diversity. Where are the wild predators? Where are the wild herbivores? Where are the trees?

Slowly and reluctantly, I have come to see that there is no good way of raising animals for food. However you farm livestock – for meat, milk, or egg production – it imposes too great an environmental load for ecosystems to absorb.

Some people respond to the twin crises of animal cruelty and environmental destruction by giving up meat, while continuing to eat fish. But commercial fishing is causing cascading ecological collapse across the blue planet.

Fishing fleets are ripping apart marine ecosystems, tearing up the complex and fragile communities of life on the seafloor, wiping out not only the target species, but turtles, dolphins, sharks, albatrosses, whales. Dead dolphins are now washing up on the coasts of Britain and western Europe in such numbers that it is hard to see how their populations can be sustained.

Here, too, we have been misled by seductive myths. When you hear the word 'fisherman', what image comes to mind? Someone who looks like Captain Birdseye: white beard, twinkly eyes, sitting on a little red boat chugging merrily across a sparkling sea?

This, again, is what our earliest children's stories show us. But it is just as far from reality as the picture-book farm. To give one example, 29 per cent of the UK's fishing quota is owned by five families, all of whom feature on the *Sunday Times* Rich List. Much of the fish people eat is now caught by vast ships, operating with gigantic nets and at great speed, scooping up everything in their path.

If we stop eating animals, we not only reduce

the cruelty humanity inflicts. We can also start restoring living systems both on land and at sea. Without livestock farming, huge tracts of land could be returned to nature. We could bring back our forests, bring back our wetlands, bring back our peat bogs, bring back our living wonders. If we stop trawlers and dredgers from ploughing the seabed, we can allow its complex living crust to reform, and the suppressed populations of fish and other animals to recover.

Not only could this shift help stop the sixth great extinction in its tracks, but it could also draw down carbon from the atmosphere, helping us to prevent runaway global heating.

We need to leave fossil fuels in the ground. We need a huge and rapid shift in the industrial economy. But we now know this is not enough. To prevent 1.5 or even 2°C of heating, we also need to extract carbon from the air. The restoration of ecosystems seems to be the most effective means of doing so.

Our aim should be to reverse agricultural sprawl, to reverse the great damage done by the fishing industry, to minimize the scale of our presence on earth, and to rewild the land and the sea. And this means retelling our stories.

Let us challenge the old myths. Let us restore our world of wonders. Let us make this the rich and beautiful world we want our children to inherit.

October 2019